Polly's Pen Pal

Mathstart®
METRICS

Polly's Pen Pal

by Stuart J. Murphy ✦ illustrated by Rémy Simard

HarperCollins Publishers

LEVEL
3

To Arnold Heynen, my in-house Canadian consultant
—S.J.M.

For Harriet
—R.S.

The publisher and author would like to thank teachers Patricia Chase, Phyllis Goldman, and Patrick Hopfensperger for their help in making the math in MathStart just right for kids.

HarperCollins®, ☕®, and MathStart®, are registered trademarks of HarperCollins Publishers. For more information about the MathStart series, write to HarperCollins Children's Books, 10 East 53rd Street, New York, NY 10022, or visit our website at www.mathstartbooks.com

Bugs incorporated in the MathStart design were painted by Jon Buller.

Library of Congress Cataloging-in-Publication Data
Murphy, Stuart J.
  Polly's pen pal / by Stuart J. Murphy ; illustrated by Rémy Simard.—
1st ed.
     p. cm.—(MathStart)
  ISBN 0-06-053168-1 — ISBN 0-06-053170-3 (pbk.)
1. Mensuration—Juvenile literature. [1. Measurement. 2. Metric system.
3. Pen pals.] I. Simard, Rémy, ill. II. Title. III. Series.
  QA465.M855 2005
  516'.15—dc22
                                                            2003027526

Typography by Elynn Cohen   13 SCP 10 9 8 7 6 5 4   First Edition

Be sure to look for all of these **MathStart** books:

To: Ally

Subject: Pen Pals

Dear Ally:

Hi! My teacher gave me your e-mail address so we can be pen pals. Cool! She said we are a lot alike even though you live in Canada and I live in the United States. We are both eight years old and we both have nicknames. My real name is Priscilla but everyone calls me Polly. And Ally is short for Allison, right? What else is the same about us? I like to play softball. What about you?

Your pal, Polly

5

"I've got a new pen pal," Polly told her family. "I wrote to her already. She lives in Canada, in Montreal. My teacher said you can learn a lot from having a pen pal in another country."

"I have a business trip to Montreal coming up soon," said Dad. "Maybe you can come with me and meet your pen pal in person!"

"Here's a message for you, Polly," said Mom the next day.

"Let me see!" said Polly.

To: Polly

Subject: Pen Pals

Dear Polly!

How awesome! I'm sure we are a LOT alike. Only I like reading about HORSES better than playing softball. My favorite color is PURPLE. What's yours?

I have one sister and one brother and one CAT. Do you have any sisters or brothers? Or pets?

I'm 125 centimeters TALL. How tall are you?

Your new friend,

Ally

P.S.: Send me your address and I'll send you a picture of myself with a note—written with a PEN!

"Is she taller than I am? How tall am I in centimeters?" Polly asked her dad later that day.

Let's see. My baseball bat is about 1 meter long. That's 100 centimeters. One quarter of that would be 25 centimeters.

Polly hurried home to write back.

To: Ally

Subject: Almost the same

Hey there, Ally:
I can't believe how much alike we are. I don't have any brothers or sisters, but I do have a cat. And guess what—my softball uniform is purple! Dad says that I'm about 125 centimeters tall, too. Just like you! How much do you weigh?

Your buddy,
Polly Romano (I'm half Italian.)

Polly checked the mail every day when she got home from softball practice. Finally a big envelope arrived from Allison Lemieux.

Polly hurried inside and ripped the envelope open.

Hello Polly,

I'm part ITALIAN, too! PLUS, I'm part DUTCH and part FRENCH. That's why I've got a French last name. I weigh just over 25 kilograms. Send me some pictures of YOU, too.

Your real PEN pal,
Ally

15

Polly ran into the backyard to show Ally's picture to her dad.

I bet I weigh about 25 kilograms, too.

Well, your weight might be a kilogram or two different than Ally's. A kilogram is 1,000 grams, and a gram is only about the weight of a leaf.

Then Dad had a surprise. "Guess what?" he said. "My trip to Montreal is next week. I wrote to the tourist office and they sent me a map. They even found a hotel for us not far from Ally's house. Do you want to come?"

"Hooray! I can meet Ally!" shouted Polly. She dashed off to write an e-mail telling Ally all about their plans.

Later, Polly and her dad looked at the map and found out that Montreal was 450 kilometers away. "How long is a kilometer?" she asked.

"Well, it's 1,000 meters," said Dad. "Let's see, we live five blocks from your school. That's about a kilometer."

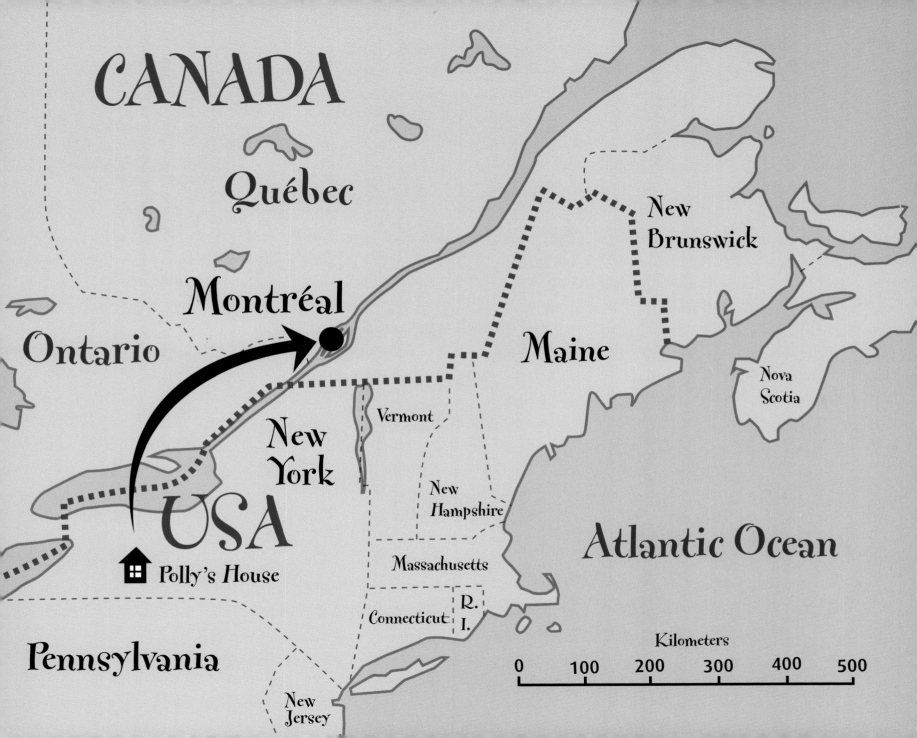

The following week, Polly and her dad were off to Montreal. Soon after they crossed the border into Canada, they stopped for gas. "We're almost on empty," said Dad. "I'll need quite a few liters."

When they got to the hotel in Montreal, Polly couldn't wait to get on the scale. She weighed 26 kilograms.

"Almost the same as Ally," she said.

Polly's dad gave her his cell phone to call Ally. "I'm here!" announced Polly.

"Great!" said Ally. "Let me give you directions to my house. Write them down."

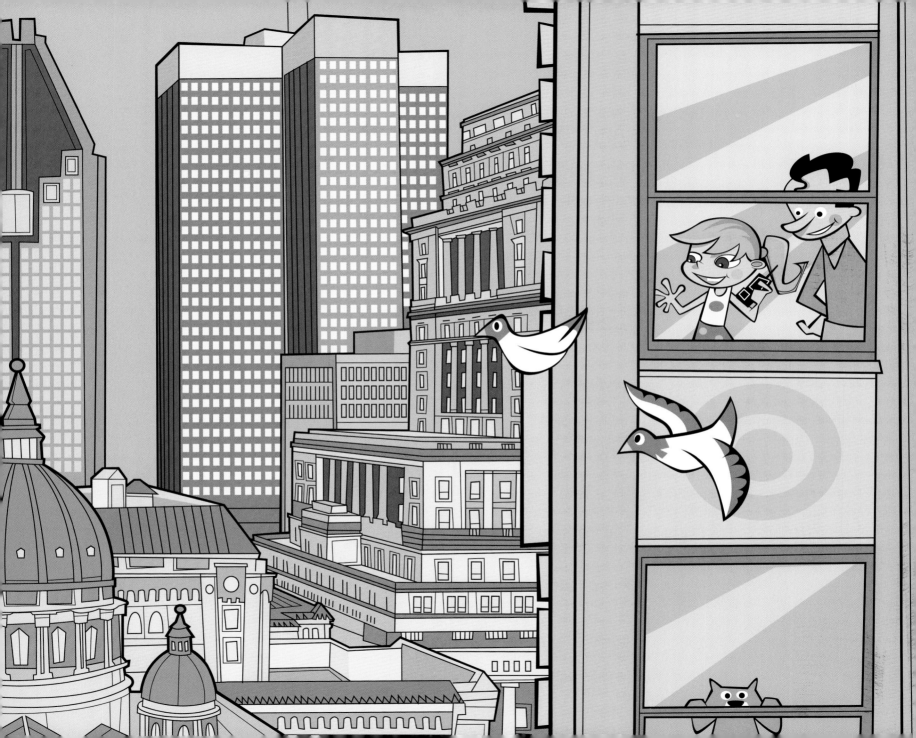

Instead of writing down the directions, Polly and her Dad hurried outside.

"First," Ally said, "turn left outside your hotel and walk about 100 meters. A meter is about two of my steps."

100 meters—that's about 200 steps.

Polly and her dad walked while Ally talked. "You'll pass
a toy store and an ice cream shop," she said. "Then you'll
be at the corner." Polly looked up and nodded.

"Turn left and walk about 60 meters," Ally went on. "You'll see a gray house. Walk up the sidewalk and you'll be at my front door."

"Okay," said Polly. "I think I've got it."

"Great," said Ally. "When do you think you can come over? Oops! The doorbell's ringing. I'll be right back."

120 more steps!

In *Polly's Pen Pal*, the math concept is measurement in the metric system. To help children become familiar with the metric system, it is important that they are given rough equivalents in terms of known objects. A centimeter, for example, might be the width of a little finger.

If you would like to have more fun with the math concepts presented in *Polly's Pen Pal*, here are a few suggestions:

• Read the story with the child and make a list of ways Polly and Ally are alike.

• In the story the length of a softball bat is used as an equivalent for a meter. Another equivalent measure might be the height of a kitchen counter. Using these as examples, help the child find other objects in the house that would be close to one meter in length or height.

• A rough equivalent for a kilogram is about the weight of a softball bat or a large grapefruit. Have the child feel the weight of one of these, and then find other objects (such as a cool iron, a sock, or a candy bar) and decide if each object weighs more or less than a kilogram.

• Using a ruler with centimeter markings, have the child measure the length or height of some familiar objects in the house.

Following are some activities that will help you extend the concepts presented in *Polly's Pen Pal* into a child's everyday life:

Metric Me: Have the child lie down on butcher paper or newspaper. Trace around his or her body with a marker. Help the child measure, in centimeters, his or her height and the length of an arm, a leg, and a little finger. Find the child's weight in kilograms. Create a chart that represents the child in metric measurements.

Take a Ride: Visit the running track at a local school. Most tracks are 400 meters around. Bike around the track two and a half times to show the child the length of one kilometer. On your next bike ride, estimate how many kilometers or what part of a kilometer you have ridden.

Metric Scavenger Hunt: Cut several strings into 10-centimeter lengths. At the child's next play date, give the children the strings and have them find objects that measure about 10 centimeters. The first child to find five objects is the winner.

The following books include some of the same concepts that are presented in *Polly's Pen Pal*:

- How Tall, How Short, How Faraway by David A. Adler
- Measuring Penny by Loreen Leedy
- Millions to Measure by David Schwartz